William Curtis

# Instructions for collecting and preserving insects, particularly moths and butterflies

William Curtis

**Instructions for collecting and preserving insects, particularly moths and butterflies**

ISBN/EAN: 9783742845016

Manufactured in Europe, USA, Canada, Australia, Japa

Cover: Foto ©berggeist007 / pixelio.de

Manufactured and distributed by brebook publishing software (www.brebook.com)

William Curtis

**Instructions for collecting and preserving insects, particularly moths and butterflies**

# INSTRUCTIONS

## FOR

## COLLECTING AND PRESERVING

# INSECTS;

### PARTICULARLY

## MOTHS AND BUTTERFLIES.

Illuftrated with a COPPER-PLATE,

On which the NETS, and other APPARATUS
neceffary for that Purpofe, are delineated.

—— *quod alii Venationibus, Confabulationibus Tefferis,
Chartis, Lufibus, Compotationibus infumunt, illud ego Ani-
makulis indagandis, colendis, contemplandis impendo.*

RAY.

## LONDON:

Printed for the AUTHOR,

And Sold by GEORGE PEARCH, Cheapfide.

MDCCLXXI.

# PREFACE.

THE following inftructions were originally drawn up for a gentleman going to refide abroad, partly with a view of collecting the various productions of nature, particularly infects.

Moft of the Englifh as well as Foreign infects, in the collections which I have lately had opportunities of obferving, have been either fpoiled in the catching, or, for want of properly know-

knowing how to preferve them, rendered imperfect, and of little or no value.

I regretted that fo much time and labour fhould be fpent to fo little purpofe; and for that reafon was induced to make thefe inftructions, with fome additions, more generally known; from which it is hoped that fome advantage may refult, not only to the young Aurelian and Entomologift, but that they may excite to a more general inveftigation of the infects of this country.

IN-

# INSTRUCTIONS

### FOR

## COLLECTING AND PRESERVING

# INSECTS.

## *Of Insects in General.*

THIS little treatise being in-
tended for those who are altogether unacquainted with insects, as
well as for those who have made them
their particular study, it may not be
amiss to premise a short account of the
nature of insects in general.

Almost every insect, from the largest
beetle to the smallest mite, is originally

pro-

produced from an egg; which the female, guided by unerring inftinct, depofits in fome place capable of affording it proper and fufficient nourifhment in its future ftate: the places to which they are committed are many and various, and ftrikingly difplay the wifdom of an all-wife Providence.

Some depofit them on the various parts of plants, fome commit them to the earth, and others to the waters; fome depofit them on putrid flefh, others in the bodies of living animals, even of infects themfelves.

After a period more or lefs fhort, the little Larvæ, Caterpillars, or Maggots, come forth; and feeding on their deftined food till they are arrived at their full growth, change into the Chryfalis or Aurelia ftate. In this they

remain

remain fome time longer ; during which they are for the moft part inactive ; and laftly come forth in their perfect or fly ftate, when they propagate their fpecies and die.

Butterflies and moths being not only one of the moft numerous, but the moft beautiful clafs of infects, whether we confider the variety or richnefs of their colours, and from the peculiar delicacy of their ftructure requiring more care to be ufed in catching as well as in preferving them, we chofe firft to fpeak of, and to be more particular in our directions concerning them.

There are two methods of collecting infects of this kind ; firft, by breeding, 2dly, by catching them in their fly ftate. Of thefe the former is much to

B 2                    be

be preferred ; as, befides the pleafure which arifes from obferving the gradual progrefs of the infect, from its egg or caterpillar to its perfect or fly ftate, we can kill them before they have in the leaft injured the farina or meal of their wings by flying.

The difficulty likewife of procuring the moft beautiful and valuable infects of this clafs in their fly ftate, makes this method much the moft eligible. Moft of the Sphinges of Linnæus, or Hawk Moths as they are called, are very feldom feen in their fly ftate, and when feen on the wing generally elude the fwifteft purfuit ; but in their cater- pillar ftate they are often found, and eafily taken. Thus the Sphinx * Atro-

pos

---

* We have ufed on this occafion the names of the celebrated Sir Charles Linnæus ; a name dear

to

pos (Jeſſamine Hawk Moth) the largeſt
and one of the moſt beautiful ſpecies
of moth this country produces, is frequently found feeding on the Jeſſamine
and Potatoe; the Sphinx Elpenor, or
Elephant Hawk Moth, on the Galium
Paluſtre, or White Ladies Bed Straw;
the Sphinx Ocellata, or Eyed Hawk,
on the Willow; Sphinx Tiliæ, or Lime
Hawk Moth, on the Lime tree; Sphinx
Liguſtri, or Privet Hawk Moth, on
the Privet; Phalæna Pavonia, or Em-
peror Moth, on the Briar, Blackthorn,
&c. and ſo of a great number of
others.

---

to every naturaliſt, and whoſe candour, genius
and induſtry can never be too much admired.
It were to be wiſhed that our Engliſh names
were in general equally expreſſive.

*Method*

## *Method of collecting them in their Cater-pillar State.*

Besides the method of collecting caterpillars by attentively examining the leaves and other parts of plants at different times of the year, there is another ; viz. by beating the boughs of trees, particularly the taller ones, with long poles, spreading a large sheet underneath to receive them.

By this means many very valuable caterpillars are often caught, which would with great difficulty be procured by any other means,

Caterpillars should be handled as little as possible ; the more hairy ones are in general the least hurt by it. It will be necessary to carry a box in the
pocket,

pocket, partly filled with leaves, to put them in.

### Method of rearing them.

The caterpillars being procured, our next endeavour is to rear them. For this purpofe they are to be fupplied with fome of the plant they are found feeding on. Although many of them live on a variety of food, the greateft part are attached to fome particular kind, deprived of which they would foon perifh.

To fave the unneceffary trouble of fupplying them with frefh food every day, feveral fprigs of the tree or plant are to be put into a wide mouth'd glafs filled with water, and the caterpillars placed on them. Moft plants may in this manner be preferved frefh for

three

three or four days. The glaſs, together with the caterpillars and their food,
is to be placed in the breeding box repreſented in the plate, and a conſtant
ſupply of freſh food is to be given them
as ſoon as the former appears in the
leaſt withered.

After they are arrived at their full
growth, they will leave off eating, and
either immediately or very ſoon change
into chryſalis ; previous to which,
Butterflies ſpin a little web, juſt ſufficient to ſuſpend themſelves by ; many
of the moths, like the ſilkworm, ſpin a
large web, in which they enwrap themſelves ; and a great number work themſelves into the earth, where they ſpin
themſelves caſes, or change without
any ſpinning ; as do moſt of the
Sphinges or Hawk Moths. It will
therefore be neceſſary to cover the bottom

tom of the box with fine mould to
the depth of three or four inches, and
keep it conſtantly moiſt.

It frequently happens that Caterpillars are what the Aurelians call
ſtung; that is, have the eggs of the
Ichneumon Fly depoſited in them: in
ſuch caſe, the eggs are generally ·
hatched before the caterpillar goes into
chryſalis. The Larvæ or Maggots
of the fly being arrived at their full
growth, eat their way out of the body
of the caterpillar, and, enwrapping
themſelves in little caſes of their own
ſpinning, change into chryſalis; and
in this ſtate have been conſidered as
the Eggs of the caterpillar, by per-
ſons unacquainted with inſects. The
caterpillar, having afforded ſubſiſtence
to its natural enemy, dies.

Cater-

Caterpillars, previous to their going into chryfalis, generally lofe the brilliancy of their colours, and many of them rove about for a confiderable time.

After remaining in their chryfalis ftate till near the time of their coming forth, fuch as are inclofed in a hard cafe or fpinning, as the Phalæna, Vinula, Pufs Moth, Phalæna Quercus, &c. are to be carefully freed from it; as the aperture which the Infect naturally makes, is often too narrow for it to pafs out without confiderably injuring its plumage. The opening will be beft made by cutting off the largeft extremity of the cafe, taking care not to wound the inclofed Pupa or Chryfalis.

*Method*

*Method of collecting them in their Chry-*
*falis State.*

Butterflies and Moths may often be
found in chryfalis under the projec-
tions of garden walls, pales, outhoufes,
in fummer-houfes, &c. and frequently
on their food.

A great variety of Moths in this
ftate may with more certainty be found
by digging in the winter months under the trees which they feed on, particularly under the Oak, Willow,
Lime, and Elm Trees. When they are
dug up in this manner, they are to be
placed as foon as convenient in a box,
fuch as is reprefented in the plate, and
kept covered with moift earth till the
enfuing fpring, when they may be dug

up

up, and placed within a few inches of
the furface of the mould, and in that
manner left to come out of themfelves.

It frequently happens, that if we
are not fo fortunate as to collect any
Chryfalis's, we are amply recompenfed
by the variety of Beetles we thus dig up.

*Method of collecting them in their Fly
State.*

The delicacy of the wings of thefe
Infects will not admit of their being
caught without injury but in Nets
made of the fineft materials.

The collector then fhould in the
firft place furnifh himfelf with a net
adapted to this purpofe: that which
is reprefented in the plate, has on
repeated

repeated trials been found to anfwer
beft. (Vide explanation of the plate.)
The net fhould be made of fine gaufe,
having its ftiffening taken off by
being foaked a little while in warm
water; or if dyed of a green colour,
which is common, this will be unneceffary.

He is next to provide himfelf with
two or three large oval boxes for the
pockets, lined at top and bottom with
thin cork; and a pin-cufhion well
ftored with pins of various fizes.

Being thus furnifhed with proper
inftruments, we fhall proceed to give
him fuch inftructions as may enable
him to ufe them with fuccefs.

It is particularly to be obferved,
that there is a continual fucceffion of
Infects

Infects, as well as of Plants. Some appear with the early Primrose, others accompany the late flowering Ivy. So that in this refpect, the Aurelian and Entomologift may regulate their excurfions by thofe of the Botanift. The latter would in vain fearch for the Pilewort ( Ranunculus Ficaria ) in the month of July; and the former be equally difappointed in feeking after the Orange Tip (Papilio Cardamines) in the month of Auguft.

Some of thofe infects continue longer in their fly ftate, and their plumage is lefs hurt by flying than others. Some continue a few days only ; others feveral weeks. In general Moths and Butterflies, unlefs they are caught the firft day of their coming out of Chryfalis, are worth little : hence arifes the neceffity of our
care-

carefully watching thofe particular
times, and of making frequent excurfions, to have them in the greateft perfection.

*Butterflies* are to be caught on the
wing only when the fun fhines warm.
They inhabit a variety of places. The
greateft number of them frequent
woods, and may be taken in or near
them; as the Papilio Iris, Populi,
Hyperantus, and moft if not all the
Fritillaries. Some delight in meadows, as the Papilio Juftina, Ga-
lathæa, Phlæas, Comma, &c. and
others frequent gardens, clover fields,
heaths, lanes, &c. Many of thofe
which frequent woods are taken with
much greater facility in the morning, a
few hours after fun-rife; at which time
they are found feeding on the flowers
that grow by the fides of the woods:
after-

afterwards, when the fun fhines with more ftrength, they fly high, and with fuch fwiftnefs as to be taken with the utmoft difficulty.

*Moths* fly chiefly in the evening, a little after fun-fet. Like Butterflies, they inhabit a variety of places, and are to be met with in the greateft plenty near woods: They may alfo be taken in great numbers in the day-time by beating the hedges, &c. more particularly in the afternoon, as the leaft motion will then put them on the wing. They are like-wife frequently met with in the day-time fticking to the bark of trees, on walls and pales that furround gardens, &c. and may be thus caught in great perfection. Some few, like the Butterflies, fly in the middle

of

of the day when the fun fhines
warm.

The ingenious naturalift, Geoffroy,
informs us that Moths may be taken
in great plenty by means of a Can-
dle and Lanthorn carried into or near
fome wood towards dark ; that the
moths immediately fly to the light,
and are caught with great facility.

We cannot recommend this method
from our own experience ; but, from
the propenfity we frequently obferve
in Moths to fly towards, and even
into, lighted candles, we apprehend
it to be a very eligible one.

An expertnefs in ufing the Net
is attained by practice only.

D                    Having

Having taken the Butterfly or Moth in the Net, we are to proceed with caution; as on killing it properly, its beauty in a great meafure depends. We are not to take hold of it indifcriminately in any part; but are, by means of the net, to bring its wings, if poffible, into an erect pofition, and then prefs the under part of the thorax or breaft betwixt the thumb and forefinger fufficiently hard to kill it: by this means the wings are neither diftorted, nor their plumage injured.

The net being then opened, the infect is to be laid hold of gently by one of the horns, and again placed betwixt the thumb and forefinger, in which fituation it is to be held while a pin, proportioned to its fize, is ftuck through the upper part of the

thorax

thorax or back. The infect is then to be placed in the pocket box.

*Method of managing them in their Fly State ; as Setting, Preferving them, &c.*

Though the infects may by this means be caught uninjured, fomething further is neceffary to make them appear to advantage. This is called fetting them, and is done in the following manner.:

The infect being ftuck through with a pin of a proper fize, is to be placed, before its wings are become ftiff, on a piece of cork having a fmooth furface, and covered with white paper. The body of the fly fhould not be made to touch the cork, when ftuck into it, but to ftand up

fome

some little distance from it. (It being only the edges of both wings that are to sit close to the cork, not the wings to lie flat on it.) The wings are then to be expanded (Vide Plate) with a fine needle, or some sharp-pointed instrument.

The upper edge of the superior wings is to be placed in a line with the head of the insect; and they are to be kept in this situation by means of little braces, made of card paper, cut in the shape represented in the Plate. These must be proportioned to the size of the wings, and fitted to their shape, by being more or less bent. By these means, the spots, &c. on both wings are made conspicuous, and the insect appears to much greater advantage.

To

To fet them well, however, re-
quires confiderable practice, and fome
ingenuity.

After remaining in this pofition
four or five days, or till it is become
thoroughly ftiff, the braces may be
taken off, and the infect removed
into the ftore box; which fhould be
fo conftructed, as effectually to feclude
thofe little infects, which are fo apt
to infeft and deftroy collections of this
kind. The fhape of the ftore box is
immaterial; that which is reprefented
in the plate is fimple, and an-
fwers every purpofe. The infide of
it fhould be lined, with thin cork;
and fome flips of cloth glued to its
edges to make it fhut clofer.

If, notwithftanding thefe precau-
tions, infects fhould get among them
(which

(which the fine powder, that falls from and furrounds fuch infects as are attacked, quickly difcovers) immerfing them in rectified fpirits of wine immediately kills them without injuring the fly. And if a little corrofive fublimate be diffolved in the fpirit ‧they are immerfed in, it will prevent them from any future attacks.

Camphire kept in the boxes is alfo a good preventative.

The flies may be either kept in boxes of this kind, or in fuch cabinets as the collector may think proper. The bottom of the drawers, if in cabinets, fhould alfo be lined with thin cork; covered with white paper,

*Me-*

*Method of collecting Insects of the Beetle Kind.*

### §. I.

By insects of the Beetle kind, we mean all such as have thin membranous wings covered with hard cases or shells, and included in Linnæus's first class of insects, COLEOPTERA. These have generally been termed Scarabæi, or Beetles: some few of them have obtained distinct English names, as the Chafer, Lady-bird, Earwig, &c. and all have been divided by Linnæus into Genera and Species.

The insects of this, as well as of the preceding and following classes, may be found in their Caterpillar or Grub state, in which they are often extremely destructive to the roots of plants ; and may in like manner be brought to their perfect or fly state, regard being had to

their

their different manner of feeding. The
time and care, however, required for
this purpofe, is probably more than can
be fpared by the generality of collec-
tors : we would neverthelefs wifh the
curious Entomologift, who has both
leifure and abilities for this purpofe, to
engage in purfuits of this kind, as
being the only means of eftablifhing
with certainty the different Genera of
Infects.

The infects of this clafs are in general eafily collected in their fly ftate.
Some creep and fly in the day-time
when the fun fhines warm ; others,
like the moths, fly in the evening and
night only.

Their habitations are exceedingly di-
verfified. Some are found in the bodies
of rotten trees, as the Lucanus Cervus,

Flying

Flying Stag, Scarabæus Cylindricus, and many of the Cerambyces : others among the dung of various animals, particularly of horfes and cows, as the Scarabæus Fimetarius, &c. Some refide in the bodies of animals that are become putrid, as the Silpha Vefpillo : great numbers are found on the leaves and ftalks of plants, as the Scarabæus Melolantha, Chafer, Coccinellæ, Lady-birds, Chryfomelæ Curculiones, &c. others delight more particularly in the flowers of plants, as the Scarabæus Auratus : fome refide altogether in woods, as many of the Cerambyces, and they are often found in great plenty under the bark of decayed trees : fome are found fwimming on the furface of ftanding waters, as the Gyrinus Natator : others in pools, ditches, ponds, &c. as the Dytifci : Some are difcovered from the light which they emit, as the

E                          Lam-

Lampyris Noctiluca Glow-worm : and, a vaſt quantity is found on dry banks, ſand banks, ſand pits, pathways, &c. particularly when the ſun ſhines warm.

Theſe inſects, as ſoon as caught, may, with a pin of a proper ſize, be ſtuck through the body, cloſe to the future that runs down the middle of the back, as repreſented in the plate (vide fig. 6) and then placed in the pocket box ; taking care that they do not injure one another from being placed too cloſe together. Or if the collector be diſpoſed to procure this claſs of inſects, he will find it very convenient, and much leſs cruel, to carry a number of ſmall pill boxes in his pockets, in which the inſects may be readily ſecured and kept till he returns home, without ſuffering any pain. They are then to be immerſed.

in

in boiling water, as being the moſt
ready means of killing them, and afterwards ſtuck through in the manner
above-mentioned ; being careful to
make the pin paſs a ſufficient length
through the body of the infeƈt, and
then placed on a piece of ſmooth cork.
When they have remained in this ſitua-
tion two, three, or four days, according to the ſize of the infeƈt, their legs,
antennæ, &c. are to be extended with
a pair of fine forceps or tweezers, and
placed in a natural poſition; in which
they will, if proper care be taken
of them, always afterwards remain.
They are then to be placed in the ſtore
box.

E 2          §. II. The

## §. II.

The next clafs of infects is the HE-
MIPTERA.

The Genera contained in this clafs
are principally thefe ; viz. *Blatta*,
Cockroach, *Mantis*, *Gryllus*, Locuft,
Grafshopper, Cricket ; *Fulgora*, *Cica-
da*, *Notonecta*, *Nepa*, and *Cimex*, or Bug.

The firft of thefe, the Blatta, like the
bed bug *, has been imported from
abroad, where they are equally numerous and troublefome. † They are
found

---

* Bugs, according to Southall, were fcarce
obferved in England till after the fire of London,
in 1666. It is fuppofed that they were then
imported with the timber which the new houfes
were built with.

† They are as troublefome as they are common
in the ifland of Senegal. Though they are fcarce
an

found in the greateſt plenty here in bake-houſes; particularly in the night, their uſual time of feeding.

All thoſe of the next genus, Mantis, are foreign ; ſome of them are extremely remarkable and curious, and from their particular ſhape have been called Walking Leaves. They are found in the meadows, and on the leaves of plants and trees. The Grylli moſtly reſide in meadows and fields among the herbage. The miſchief done by the Blattæ is nothing, compared with the ravages of ſome of thoſe,

viz.

---

an inch thick, they do an incredible deal of miſchief. They gnaw linen, ſheets, wood, paper, books, and in ſhort whatever comes in their way : they attack even the Aloes, the bitterneſs of which keeps off all the other inſects.

Adanſon's Voyage to Senegal, p. 296.

viz. the Locuſts.* One ſpecies of this genus, the Gryllus Domeſticus, reſides in houſes, particularly where there are ovens.

* In this voyage (ſays Adanſon) I was witneſs myſelf, for the firſt time, to the miſchief done by Locuſts; that ſcourge ſo dreadful to hot climates! The third day after our arrival we were ſtill in the road, when there ſuddenly aroſe over our heads, towards eight o'clock in the morning, a thick cloud, which darkened the air, and deprived us of the rays of the Sun. Every body was ſurpriſed at ſo ſudden a change in the ſky, which is ſeldom overcaſt in this ſeaſon: but we ſoon found that it was owing to a cloud of Locuſts, raiſed about twenty or thirty fathoms from the ground, and covering an extent of ſeveral leagues; upon which it poured a ſhower of thoſe inſects, which fell to devouring while they reſted themſelves, and then reſumed their flight. This cloud was brought by a very ſtrong Eaſt wind; it was all the morning in paſſing over the adjacent country; and we imagined that the ſame wind drove the Locuſts into the ſea. They ſpread deſolation wherever they came; after devouring the herbage, with the fruits and leaves of trees,

ovens. Moſt of the Fulgoræ are diſcoverable from the light which they emit; theſe, like the Mantes, are foreign, and many of them equally curious. The Cicadæ are found on trees and plants; the Notonectæ and Nepæ reſide in ſtagnating waters. There is ſcarce a perſon who has lived a while in any very populous place, but knows where

---

trees, they attacked even the buds and the very bark : they did not ſo much as ſpare the reeds with which the huts were thatched, notwithſtanding that theſe were ſo dry : in ſhort they did all the miſchief that can be dreaded from ſo voracious an inſect.

The inhabitants of Aſia, as well as Europe, ſometimes take the field againſt Locuſts, with all the dreadful apparatus of war. The Baſhaw of Tripoli in Syria ſome years ago raiſed 4000 ſoldiers againſt theſe inſects, and ordered thoſe to be hanged who refuſed to go.

Haſſelquiſt's Voyage to the Levant, &c.

where to find one species of the next genus, the Cimex or Bug, to particularize the places where the remaining species of this very numerous genus reside, would be almost endless; they are in short to be met with almost everywhere.

These insects may be killed either with boiling water, or a few drops of spirit of turpentine. They are all of them to be stuck through the thorax or back betwixt the shoulders. The wings of the Grylli, and some of the others, are to be expanded, and kept so by means of the little braces; and their legs, antennæ, &c. are also to be placed in a natural situation.

§. III. The

## §. III.

The infects contained in the next clafs, NEUROPTERA, are chiefly aqua- tic, refiding in the waters in their caterpillar ftate, and flying about them in their perfect ftate. The principal genera are, the Libellula, Dragon Fly, Ephemera, May Fly, Phryganea, He- merobius, and Panorpa. The Libel- lulæ are confidered by the generality of people as containing in them fome- thing venemous, and from hence have derived their feveral names of Adder Spear, Adder Bolt, Horfe-ftinger, &c. It muft be confeffed, that their fhape, manner of flight, &c. is fuch as might readily raife fuch an idea : the collector, however, muft not be mifled by their appearance, and intimidated from catch- ing them, they being perfectly harm-

F            lefs,

lefs, indeed equally fo with the Gnats
they feed on.

The net ufed for catching Butterflies
will be very convenient for catching
the infects of this clafs, particularly the
Libellulæ.

They are all of them readily killed,
either by fqueezing their thorax, or
with a few drops of fpirit of turpentine.
The fame means are to be ufed in
fetting them, &c. as in the Hemip-
tera.

§. IV.

The infects of the next clafs, Hy-
MENOPTERA, are for the moft part
armed with ftings, which are either
venemous or harmlefs. The principal
genera are the Tenthredo, Ichneumon;
Sphex,

Sphex; Chryſis ; Veſpa, *Waſp*; *Hor-
net*; Apis, *Bee*; Formica, *Ant.*

The Tenthredines are found on trees
and flowers in their caterpillar ſtate :
they feed on the leaves of plants. The
Ichneumones are found in the ſame
manner : in their caterpillar ſtate they
live chiefly in the bodies of other inſects, particularly of the caterpillars of
Moths and Butterflies, in which they
depoſit their eggs. The Sphex reſides
principally in ſand banks ; it is alſo
caught on flowers, &c. This inſect
catches and kills others, which it bu-
ries in the ſand, having previouſly
depoſited its eggs in them. The
Chryſis, (many ſpecies of which are
exceeding beautiful) is found flying
about old walls, poſts, ſand banks,
&c. in which it builds its neſt.
*Waſps, Bees,* and *Ants* are found on

flowers

flowers and fruits, and almoſt on every thing that is ſweet.

Theſe inſects being armed many of them with poiſonous ſtings, it will be neceſſary to uſe the forcep nets to catch them with (vide fig. 7.) When caught, a pin is to be ſtuck through them while in the net (i. e. through the thorax, as in fig. 8.) It is very difficult to kill theſe inſects without injuring them in ſome reſpect ; boiling water hurts their wing, and the fine hairs with which the bodies of many of them are covered ; ſpirits of wine or turpentine proves immediately fatal to ſome, while others are ſcarce affected by it ; and letting them remain transfixed till they are dead, will probably be thought too cruel.* We ſhall leave the collec

tor

* The beſt method, as I have ſince been informed by an ingenious gentleman, is to ſtick

them

tor to adopt either of thefe methods, or any other he pleafes. When dead, their wings, &c. are to be expanded, and kept in as natural a pofition as poffible.

## §. V.

Infects with two wings only, form the next clafs, DIPTERA : it contains various kinds of *flies* and *gnats*. There is fcarce any place in which fome of the former are not to be met with ; but they are found more particularly on all kinds of plants and flowers, efpecially on the umbelliferous ones. Some of them fly about cattle of various kind, in the fkins of which they depofit their eggs, as the Oeftrus Bovis, &c. The latter

moftly

---

them through with a needle dipt in aqua-fortis. The Sphinges, and other large Moths, are like-wife killed in this manner with the leaft injury.

moſtly fly about waters and watery places.

Theſe inſects are eaſily killed by a few drops of ſpirit of turpentine : their wings are to be expanded ſo as that their bodies may become apparent ; and a little brace ſhould be placed underneath them, to prevent their bodies from being too much incurvated in drying, which they are very apt to be.

## §. VI.

The inſects of the laſt claſs are ſuch as have no wings at all, and are therefore called, APTERA. *Spiders, Scorpions, Centipes,* and *Crabs* of various kinds, make up the principal part of this claſs. Theſe are ſo common, and the places they inhabit ſo generally known, that any information on the

means

means of collecting them would be
fuperfluous.

Moft of them require to be preferved
in fome kind of fpirits, as fpirits of
wine, rum, brandy, &c. Such of them
indeed as are inclofed in hard fhells,
may be preferved dry, in the fame
manner as the infects of the Beetle
kind.

As the collector will have frequent
occafion for the ufe of cork, both to
line his boxes with, and to fet his infects
on, we fhall informhim how to prepare
it for ufe.

He may procure the cork in large
pieces at any of the cork-cutters.
Thefe muft be cut into fmaller ones;
and in order to make the cork flat,
it is held before the fire till it becomes
hot

hot through; it is then to be immediately placed betwixt two fmooth
boards, and a very heavy weight laid
on it, where it muft remain till cold.
Thus flattened, it is to be rafped on
both fides, with fuch a rafp as is ufed
by the bakers; afterwards with a
finer one; and laftly with a pumice
ftone, which makes it perfectly
fmooth. If the cork be thick, and
the purpofe of it to line boxes, it may
be fawed through the middle, and
rafped on both fides as before.

EXPLA-

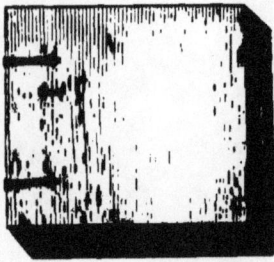

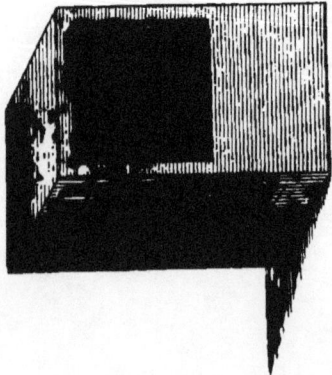

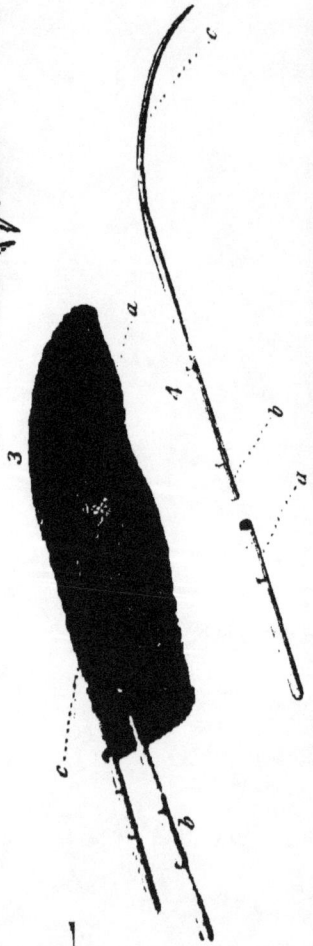

---

# EXPLANATION

## OF THE

# PLATE.

Fig. 1. The breeding box. *a.* an opening in the front, covered with gaufe ; *b.* the door on the fide.

N. B. The bottom of it fhould be covered with fine mould to the depth of four or five inches. The fhape and ftru&ure of it need not be ftri&ly ad-hered to.

2. Shews the manner in which the wings of the Butterflies and Moths

G                              are

are to be expanded, and how the braces are applied to keep them so.

3. The Net used for catching Moths, Butterflies, and other insects. *a.* the part made of fine gause; *b.* the sticks; *c.* the binding which receives them.

N. B. The sticks, for the conve‑ nience of carrying them, are made to take to pieces, somewhat in the man‑ ner of fishing rods. They join to one another by means of hollow brass ferrils fixed to the end of each. There are three of them, each of which is about fourteen inches long.

4. Shews one of the sticks; *a.* the brass ferril; *b.* the end of the next stick, which goes tight into it.

To

To the upper end of the fticks is
joined in like manner a piece of cane,
about two foot long, bent of a proper fhape, *c.* Inftead of three pieces
of wood here reprefented, the other
ftick may confift of one entire piece,
and be ufed as a walking-ftick. The
gaufe muft be edged with two pieces
of binding fewed together to receive
the fticks ; and as the fticks are taper,
fo muft be the cavity : at the upper
part of it, where the fticks meet, it
muft be clofed by a few ftitches, that
the net fticks may fhut even together.
The net may be about a yard broad
when expanded, and the length of it
a yard and a quarter. This fize,
however, may be varied at pleafure.

5. The Store Box.

6. Shews

6. Shews the manner in which the Beetles, &c. are ftuck through one of the wings.

7. The Forcep Nets.

8, Shews a fly ftuck through the thorax: and in the fame manner all the different fpecies of Bugs, Wafps, and Flies in Sect. II. III. IV. and V. are to be ftuck.

F I N I S.

---

*Speedily will be publifhed,*

A Tranflation of the FUNDAMENTA ENTO-MOLOGIÆ of Linnæus; or, " An Intro-duction to the Knowledge of Infects."